Cyber Essentials

A guide to the Cyber Essentials and Cyber Essentials Plus certifications

Cyber Essentials

A guide to the Cyber Essentials and Cyber Essentials Plus certifications

ALAN CALDER

IT Governance Publishing

IT Governance Publishing Ltd
Unit 3, Clive Court
Bartholomew's Walk
Cambridgeshire Business Park
Ely, Cambridgeshire
CB7 4EA
United Kingdom
www.itgovernancepublishing.co.uk

First edition published in the United Kingdom in 2023 by IT Governance Publishing.

ISBN 978-1-78778-434-5

Cover image originally sourced from Shutterstock®.

ABOUT THE AUTHOR

Alan Calder is a leading author on IT governance and information security issues. He is the CEO of GRC International Group plc, the AIM-listed company that owns IT Governance Ltd.

Alan is an acknowledged international cyber security guru. He has been involved in the development of a wide range of information security management training courses that have been accredited by the International Board for IT Governance Qualifications (IBITGQ).

He is a frequent media commentator on information security and IT governance issues, and has contributed articles and expert comment to a wide range of trade, national and online news outlets.

ACKNOWLEDGEMENTS

I would like to thank Nigel Evans, Technical Writer at GRC International Group Plc, for his help developing the material in this book.

CONTENTS

Contents

CHAPTER 1: THE CYBER ESSENTIALS SCHEME

Cyber attacks are a fact of life in the information age. For any organisation that connects to the Internet, the issue is not if an attack will come, but when. Most cyber attacks are performed by relatively unskilled criminals using tools available online. These attacks are often opportunistic: looking for easy targets rather than rich pickings.

The Cyber Essentials scheme is a UK government-backed effort to encourage UK-based organisations to improve their cyber security by adopting measures (called controls) that defend against common, less-sophisticated cyber attacks. The scheme recommends practical defences that should be within the capability of any organisation. Cyber Essentials is the digital equivalent of a locked front door and closed windows, rather than barbed wire, guard patrols and watchtowers.

The Cyber Essentials scheme was created in 2014 by the National Cyber Security Centre (NCSC), which is a part of the UK government. There was a major update to the scheme in 2022, which changed some of the requirements, and a smaller update in January 2023. These changes were largely prompted by developments in technology and working practices in the years between: most of the changes deal with Cloud-based services, bring your own device (BYOD), home working and authentication methods (including modern password methodologies and multi-factor authentication). The requirements of the scheme are available on NCSC's website.[1]

The Cyber Essentials scheme has two levels: the basic Cyber Essentials and Cyber Essentials Plus. An organisation can be certified to either level. The difference between the two levels is not in security precautions: the two have the same requirements in terms of security controls and, in theory, provide the same level of security. The difference between the two levels is in assurance: the basic scheme is assured by a questionnaire filled in by the applicant organisation. Cyber Essentials Plus mandates a device test from an external certification body and a scan of the organisation's infrastructure to reveal any unaddressed vulnerabilities.

The scheme is certified by NCSC's partner, the IASME Consortium. IASME regulates certification bodies, the companies that provide certification for Cyber Essentials.

Why get certified?

Cyber attacks are something that every organisation needs to take seriously. The most important reason to become certified to Cyber Essentials is to get the cyber security of your organisation in order, and to protect your data, systems and staff from expensive and disruptive cyber attacks.

There is a significant chance that you are reading this book because Cyber Essentials certification is required to bid for certain UK Government contracts. These contracts can be

---

[1] *https://www.ncsc.gov.uk/cyberessentials/resources*.

very lucrative and your company may be willing to deal with a lot of red tape to get one. Cyber Essentials should not be seen as a box-ticking exercise or a bureaucratic annoyance, however. The scheme requires simple, practical measures to improve security and the scheme's benefits are likely to outweigh the modest commitment in effort that its controls require.

Cyber security is important to private companies as well as government organisations. In today's climate, the business case for certification to a scheme like this goes beyond obtaining government contracts. Both customers and vendors are likely to be looking to you to keep data safe. Cyber Essentials certification represents an organisation's commitment to at least a solid foundation of cyber security.

Assuming that your organisation isn't a total newcomer to the field of cyber security, there is also a good chance that you have already implemented many of the controls, so becoming certified is not only valuable but often quite easy. None of the controls are particularly onerous or frivolous.

This is no reason for complacency, however; even large organisations may not have covered every control. To ensure that your ability to bid for a contract is not undermined, to protect from future legal consequences and to make sure that you only go through the auditing process once, it is crucial that you ensure you are fully compliant with all the Cyber Essentials technical control themes. This book should help you to achieve that goal.

Which contracts require Cyber Essentials?

In general, any organisation bidding on a contract to supply UK central government that involves data processing and/or IT services is going to require Cyber Essentials certification (the basic level) as a minimum.[2] Contracts that involve large or sensitive data sets, or are let by the Ministry of Defence, are likely to require Cyber Essentials Plus certification.

The above is government purchasing policy, but certification can be required in other circumstances, at the option of the organisation letting the contract. For example, UK local government and other government departments may specify that their data services can only be provided by those companies certified to Cyber Essentials or Cyber Essentials Plus. Private companies are also free to require organisations to be certified before bidding for contracts.

Note that Cyber Essentials certification may not be necessary for organisations that are certified to a more expansive and rigorous cyber security standard. For example, if an organisation is certified to ISO 27001, which is much more demanding than Cyber Essentials, it will not need a separate Cyber Essentials certification, if the body

---

[2] *https://assets.publishing.service.gov.uk/government/uploads/system/uploads/attachment_data/file/526200/ppn_update_cyber_essentials_0914.pdf.*

providing its ISO 27001 certificate is also authorised to certify Cyber Essentials applications.[3]

What am I protecting?

In terms of your organisation's infrastructure, you need to protect any device that can connect to the Internet (directly or via another device), any software that runs on these devices and any data that the devices store or process. The devices you need to protect include desktop machines, laptops, servers, firewalls, routers and mobile devices. You must take attacks on all your infrastructure seriously. Low-skill cyber attacks are the most common and these are, in practice, targeted at the most vulnerable elements of your IT infrastructure. This is because these attacks are often sprayed in high volume across the Internet with little thought as to a target. Any attack that succeeds at finding a vulnerability is given extra attention by the criminal. This means that your infrastructure is likely to be subject to many low-level attacks, and the sheer volume of these is likely to find any serious gap in your defences.

The aim of most cyber attackers is to steal data such as sensitive business information or financial records. The personal details of staff and customers are a common target, and if you have access to data that can be used for the purposes of fraud (such as payment card data) your organisation will be of particular interest to online criminals. Passwords are also valuable, especially because

---

[3] *https://assets.publishing.service.gov.uk/government/uploads/system/uploads/attachment_data/file/526200/ppn_update_cyber_essentials_0914.pdf*.

of 'credential stuffing' – a type of attack that relies on the tendency of people to reuse their passwords. This habit means that if a password is discovered on one system, an attacker can use it for accounts held by that user on different systems. Capturing a large number of passwords is therefore a great treasure for criminals.

There are other reasons to perform cyber attacks. Ransomware, where a criminal cripples an organisation's systems in the hope of extorting a cash ransom, has risen from a rare annoyance to a flourishing industry. It is also possible that it's your computing power that a hacker wants: they may wish to turn your devices into 'zombie' machines under their control, to increase the reach of their other attacks or otherwise take advantage of your networks.

Cyber attacks are a common experience for small businesses: 42% of small businesses suffered a successful cyber attack in 2021.[4] The security measures required by Cyber Essentials are sensible, practical measures that are good practice for even small organisations. Failure to take cyber security seriously can result in theft, fraud and even legal repercussions – in other words, by putting these controls in place, you are defending critical areas of your business. You are also protecting your reputation – it is highly embarrassing to publicly admit that you have been the victim of a low-tech cyber attack, because it shows all your customers that you cannot protect your information and are unlikely to protect anything they entrust you with.

---

[4] *https://advisorsmith.com/data/small-business-cybersecurity-statistics/*.

Beyond and outside Cyber Essentials

It is worth noting that the Cyber Essentials scheme only lays out a security approach that ensures a basic level of protection against unsophisticated threats. As such, the controls discussed here are just the starting point for companies that are serious about protecting themselves, their data and their customers. Organisations facing more advanced or determined opposition – especially attacks targeted specifically against the organisation – should create a stronger security position. Fortunately, the security requirements laid out by Cyber Essentials are in line with well-established standards such as ISO 27001. The Cyber Essentials controls often form a subset of the requirements for these advanced standards and therefore implementing them can form the solid foundation of a more comprehensive security infrastructure as your organisation's capability evolves.

Note that since it has been designed to cover only the most common software and hardware systems in use, certain varieties of software cannot be certified as secure under Cyber Essentials – for example, point of sales (POS) software, PIN-entry devices (PED) and e-commerce applications. These systems have different vulnerabilities and therefore require some different kinds of protection on top of the basic rules outlined in the scheme. Organisations that make significant use of these might have to also look at standards that deal with such items, like the PCI DSS.

Structure of this book

This first part of this book will examine the various threats that are most significant in the modern digital environment, their targets and their impacts. It will also cover implementation and documentation issues. We will then discuss the Cyber Essentials requirements for infrastructure in detail, following the scheme's own organisation of five technical control themes:

1. Firewalls
2. Secure configuration
3. User access control
4. Malware protection
5. Security update management

We will explain why these controls are useful and suggest strategies for implementation.

Secondly, we will look certification. Most organisations with an interest in Cyber Essentials will want to become certified, whether at the basic or Plus level. We will discuss scoping and ask questions that will help you determine whether your organisation is ready for certification. We also go into detail on how the process for certification works, for both levels of the scheme.

The final part of the book presents a selection of additional resources that are available to help you implement the controls or become certified: it includes further reading and consultancy options, and also covers some sensible steps your organisation can make if you would like to take cyber security beyond the basic level mandated in Cyber Essentials.

Part 1: Requirements for basic technical protection from cyber attacks

CHAPTER 2: TYPES OF ATTACK

The controls set out in the Cyber Essentials requirements document[5] will provide security to organisations of all sizes but have been chosen because they are relatively easy to implement for smaller organisations, and they protect against a wide variety of common cyber threats. But what are the attacks that your organisation faces?

The image of the hacker in popular media is usually of a lone individual in a basement, tapping away at a keyboard, trying to break into a specific computer system. This is no longer how most hackers work. Hacking is a business and cyber criminals are interested in getting efficient returns. A targeted attack methodology is not often efficient, which is lucky because it is difficult to keep out a motivated and expert cyber criminal who is deliberately targeting your organisation.

Successful cyber attacks generally rely on simple technology that is widely available on the Internet. Many hackers are not particularly skilled and they concentrate on getting modest returns quickly. The efficient way to perform cyber crime is to launch an attack at many organisations at once and then concentrate on those whose defences are easily breached. If an organisation's front door is unlocked, then they are unlikely to have effective guard dogs and burglar alarms. So an attacker will look for

---

[5] *https://www.ncsc.gov.uk/files/Cyber-Essentials-Requirements-for-Infrastructure-v3-1-January-2023.pdf.*

poor technical controls or unaware staff in the first instance.

Setting up basic defences will not deter a skilled, determined attacker. However, most attacks will be repulsed by even modest measures and these are the defences that Cyber Essentials mandates.

Here are the attack types that Cyber Essentials seeks to protect against:

Social engineering

The most vulnerable part of an organisation is generally staff rather than technology. Social engineering attacks seek to manipulate an organisation's staff (and people in similar roles, such as contractors and vendors) into performing unwise actions that benefit the criminal. Social engineering attacks often appear via email, such as a spam email with a link that unleashes a virus when clicked. Although email is the most popular delivery method – it is cheap and easy to spoof – other methods exist: putting a key logger on handfuls of USB drives and scattering them in an organisation's car park in the hope someone will pick one up and plug it into a work machine is a social engineering attack.

Phishing is an increasingly important sort of social engineering attack. This is where the criminal impersonates someone of importance or authority and urgently requests an action or information. This is often a request for credentials (e.g. sign in to a fraudulent website) or to send money or information to a spoofed address that looks authentic but goes to the criminal. Again, email is the most popular method, but others exist, such as SMS text scams.

Denial of service (DoS)

This sort of attack seeks to stop something working, often a website or other Internet-based service. The attacker generates a huge number of spurious requests to the service, such as a million requests per second to view a website. The service cannot cope with this many requests and shuts down. A variant of this is the distributed denial of service attack (DDoS), which uses a host of computers (usually compromised by malware) to increase the scale of the attack and makes it harder to block. While the measures specified by Cyber Essentials can help against becoming a victim of a DoS attack, their main benefit in this area is stopping devices being compromised by malware and turned into 'zombie' machines – tools for criminals.

Password attacks

Passwords are the ubiquitous method for authentication to computing services, and there are many ways in which a criminal can guess or obtain a password. The least elegant is brute force: attempting to enter every possible combination of password characters until a match is found. The job is much easier for an attacker if a weak or common password is used, as brute-force attacks will usually start by checking these. However, brute-force attacks are a lot of work and are easily foiled, and there are other options. Key loggers are a type of malware that record keyboard inputs and pass them to another party. No rules for a lengthy or complex password will stop these: the logger must be detected and disabled before a password is typed in.

Credential stuffing is an attack type that exploits the tendency to use the same password or small set of passwords across multiple sites and services. If an attacker compromises a user's password on one service (or buys a list of compromised passwords from another criminal), they may try the same password on other services that a user has an account for. In this way, a user's account with a service can be compromised through no fault of the service: it is the user reusing passwords that caused the breach. This is a particular problem if a user reuses passwords between personal and work accounts.

Threats outside the perimeter

An organisation will typically have computing assets on a network or networks, all surrounded by a well-defined border guarded by perimeter devices such as firewalls. There was a time when this was commonly the whole of an organisation's architecture, but this is rarely still the case. Organisations increasingly rely on Cloud-based services and fundamentally that means using another organisation's computers. This, and the rise of remote working and bring your own device (BYOD), means that many of the devices that access an organisation's data are outside the security perimeter.

Cloud-based services are subject to the same attacks as any other networked system, such as social engineering or password attacks. They require particular attention because many of their security controls are under the control of another party, and having your own security posture in order may not help if your Cloud service provider leaves holes in its defences. This is why all Cloud-based services used by an organisation must be in scope for Cyber

Essentials. Another issue is that Cloud-based computing naturally entails passing data out of your own network and across the Internet. Interception is therefore a risk and encryption needs to be considered.

Remote working and BYOD means that many devices that are not controlled by an organisation can be used to access its data. This leaves organisations vulnerable to attacks on misconfigured devices, and to man in the middle (MITM) attacks, where an attacker listens in on an unsecured connection (such as a coffee shop wireless network).

Misconfiguration and unpatched vulnerabilities

Modern devices and software are complex and usually need to be configured. For example, a firewall is of little use unless the rules that govern what connections it allows or stops are chosen correctly. Many devices are a potential entry point for attacks, so need to be configured to minimise the risk. This often entails removing unnecessary accounts or functionality from the device.

Software often has vulnerabilities, even when correctly configured. New vulnerabilities in software are found daily, by both criminals and legitimate security professionals, and software vendors rush to close these gaps in security by releasing patches to their software. This creates a race between criminals and IT managers. A patch is of no use until it is applied to the software, so it is important to apply patches that address security issues swiftly or automatically.

Ransomware

Ransomware attacks have boomed to become an industry in recent years. An attacker inserts malware into an organisation's network, which denies the organisation access to its own data and threatens to publish, delete or permanently encrypt it unless a ransom is paid. Paying a ransom is expensive, may be illegal and in no way guarantees that the data will be recovered safely – in fact, it is very unusual for a ransom payment to result in complete data recovery. Ransomware can enter an organisation's networks through vectors such as social engineering: it can take just one click on a malware-bearing link to infect a large organisation. Cleaning up after a ransomware attack is often a laborious, expensive process, whether or not the ransom has been paid.

Scoping

An important early step in engaging with the Cyber Essentials scheme is to understand your organisation's infrastructure and decide what will be in scope for certification. The scope must be agreed with the certifying authority before certification can begin. In general, you should choose the whole of your IT estate to be in scope. This will provide the greatest protection to you and the data you hold. It will also provide the greatest assurance that your organisation is fully committed to the scheme and the security it brings.

If necessary, however, you may wish to include only a well-defined sub-set of your infrastructure. This is done in terms of the networks the sub-set uses. If you do limit the scope, it must be clear who manages the sub-set, what equipment it uses and what data it processes. Scopes can be defined as named networks, or as the whole organisation excluding certain networks. Infrastructure that deals with sensitive information should be in scope. Any sub-set must be segregated from the organisation's other networks (e.g. by firewalls or an air gap).

Examples of good scoping include:

- The whole company, excluding a specific scientific network.
- The whole UK company, excluding workshop, test and guest networks.

Examples of bad scoping include:

- The corporate environment.
- All systems and infrastructure that process customer data.
- The marketing team.

The 2022 version of Cyber Essentials expanded the requirements for items that need to be in scope. For example, Cloud-based services must now be within scope. BYOD hardware is in scope if it accesses organisational data, and some home working equipment will also be in scope of the scheme.

This book examines scoping in more detail in Part 2.

Implementation and documentation

Before you start implementing the controls that Cyber Essentials requires, you should establish an approach to documenting your progress that can be used with all five measures. Most of the documentation should consist of policies and procedures that set out how you are implementing the controls and what your staff need to do.

It is important to document all the controls as you plan and implement them. Documenting the controls will help you implement them consistently. You may also need the documentation for compliance purposes, including filling in the self-assessment questionnaire when the time comes for Cyber Essentials certification.

Your documentation should be based on the Cyber Essentials controls and explicitly reference the network and user devices that are in scope. It should be easily accessible to staff and form part of your IT security policies and procedures.

CHAPTER 3: TECHNICAL CONTROL THEMES

To become certified for either Cyber Essentials or Cyber Essentials Plus, you must implement various security measures, called controls, to ensure the safety of your systems and data. Controls are often technical in nature (e.g. automatic application of new security patches) but can also be based on policy (e.g. a review of user accounts to see they are still necessary) or people (e.g. ensuring staff are educated on what makes a good password).

The controls are designed to provide defences against the basic attack types described previously and are most effective against untargeted or low-skill attacks. Although implementing these controls is a prudent measure and provides a sturdy basic defence, doing so is a first rather than a last step on the road of information security.

The controls of the Cyber Essentials scheme are organised into five technical control themes as follows:

1. Firewalls
2. Secure configuration
3. User access control
4. Malware protection
5. Security update management

The detailed list of technical controls required by the Cyber Essentials scheme is available from the NCSC web site[6].

Technical control theme 1: Firewalls

Although the term 'firewalls' often refers to specific hardware devices, the Cyber Essentials scheme includes any device or software that manages the flow of traffic via data flow rules. This means that the scope of the firewall requirements includes:

- Laptop and desktop computers;
- Traditional boundary firewalls;
- Routers;
- Servers;
- Cloud-based services; and
- Other services delivered over the Internet.

The objective of this technical control theme is to limit the services that can be accessed from the Internet to those that are both necessary and safe.

Devices connected to the Internet can access various services provided by other machines, whether this is a remote worker downloading a file from shared storage or someone in an office replying to an email. If a device is inside a security boundary, there should be a firewall on the boundary that controls traffic. The firewall will have rules

---

[6] *https://www.ncsc.gov.uk/files/Cyber-Essentials-Requirements-for-Infrastructure-v3-1-January-2023.pdf.*

that define which traffic is to be allowed past and which is to be blocked.

If the device is not inside a boundary (such as a home worker's laptop), then the device itself should have a software-based (sometimes called 'host-based') firewall that performs the same function. Software-based firewalls mean that the rules apply to a device wherever it is used. Their rules also be configured at the level of the individual device, which allows a more granular and tailored approach to security, albeit at the cost of an increased administration burden. Another advantage of software firewalls is that they can stop malware from infecting a device even if it has penetrated the organisation's security boundary.

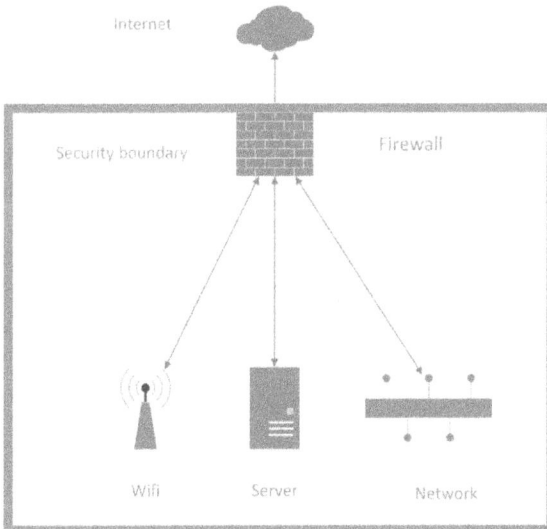

Figure 1: A firewall on an organisation's security boundary

Restricting the traffic that a device accepts reduces the number of targets than an attacker can exploit and makes the device safer. Cyber Essentials requires that every device in scope is protected by an appropriately configured firewall, or equivalent device or software. Software-based firewalls must be installed and configured on any devices in addition to boundary firewalls.

As firewalls are security devices that control access to a network or machine, they are themselves valuable targets for an attacker and must be appropriately secured. Firewalls are generally protected by local administration credentials, which need to be kept secure. A common basic error with firewalls is failing to change the default administrator password to a new, strong password (see the later section on passwords for details). The default passwords are often available on the Internet and failing to change them leaves an open door for an attacker.

Firewalls are operated by an administrator console. These are available when physically at the device but can often be used remotely across the Internet. While convenient for firewall administrators, this is not secure. Internet-based access to firewall administration should be disabled as a general practice. Internet-based firewall administration should only be allowed if there is a clear and documented business need for it and the interface is protected by additional measures beyond a simple password (such as multi-factor authentication or a whitelist of a small number of approved addresses that the console can be used from).

The heart of a firewall is its set of rules: a firewall with rules that permit all traffic may as well not be there at all. The rules of each firewall must be configured appropriately to block dangerous or unnecessary traffic. The details of this will depend on the business needs and security requirements of an organisation, but Cyber Essentials mandates that inbound connections that cannot be authenticated should be blocked by default and the rules of an organisation's firewalls should support this.

Newly purchased firewalls often have a set of default rules, and changes to an organisation's business or infrastructure mean that firewall rules need to change. These two factors mean that the rules for a firewall can become out of date. Cyber Essentials mandates that firewall rules are kept in review and rules that are no longer needed are quickly removed or disabled. This reduces the risk of an attacker finding a vulnerability in an old rule that the firewall administrators have forgotten.

Technical control theme 2: Secure configuration

Both hardware devices and software can generally be configured in many ways. Secure configuration has two main goals:

1. Ensure that the vulnerabilities in software and devices are minimised to the greatest extent that is practical without impacting business requirements.
2. Reduce the attack surface available to a hacker by removing or disabling any features or services that are not required for business use.

Although secure configuration is a matter of ongoing management, the moment when a new device is first switched on and added to an organisation's infrastructure is critical. A newly-installed device may have several problematical configuration issues, such as:

- A default administrative account with significant privileges, which may have a widely known or weak default password;
- Services or functions that are not necessary for the role the device has in the infrastructure; and/or
- Default user accounts, some of which may have increased privileges.

For this technical control theme, Cyber Essentials stipulates that an organisation must actively manage the configuration of its devices. It is good practice to create a standard configuration for each type of device (e.g. staff laptop or boundary firewall) and apply that configuration to each device of that type. These standard configurations are known as 'images'. This will reduce the burden of device management and the risk of 'rogue' unmanaged devices. It also means that staff can more easily change devices, as they will already know where the relevant software is and won't have to deal with a strange interface. Furthermore, your technical support staff can more easily implement common security controls, manage patches and vulnerabilities, and identify when a device is not functioning correctly

All unnecessary accounts – both those created before purchase and those created for staff but no longer needed – must be removed or disabled to prevent them being misused by hackers. Your system is only as secure as the most vulnerable account, so if a default account is simple to break into then it is a threat to your entire IT

Infrastructure. Default accounts are easy to forget about, and it is important that an organisation's security policies manage them competently.

Similarly, all unnecessary software must be removed or disabled, including applications, system utilities and network services. Some software can provide hackers with access to sensitive information, while unremoved communications software (e.g. a messaging service) could be hijacked to send data outside the company undetected. Even when a piece of software has no obvious vulnerabilities, it is safer to remove or disable it if it is not required for business purposes.

If a default account is not removed, then the password to it must be changed. The new password must obey the organisation's general password policies (see later for details of the requirements for these). Default passwords may well be weak or available across the Internet, whether in a manufacturer's documentation or distributed by attackers. This issue is especially important for highly privileged (e.g. administrator) accounts. A hijacked privileged account can add or delete other accounts, making it much easier for an attacker to cause significant damage or operate undetected.

A danger when downloading files from the Internet is auto-running software: software that executes itself without action from a user. This allows free reign for any malware that enters an organisation's infrastructure and stops staff from intercepting it. Properly educated staff are likely to at least pause when there is a suspicious email with an executable attachment, but that education will be for nothing if the malware in the attachment can run itself. All auto-run features should be disabled on all devices within an organisation's scope. This will also stop attacks such as scattering USB sticks with auto-running malware on them in the staff car park in the hope that someone will plug one into a work machine.

All users must be authenticated before they use any of the organisation's data or services. This is an obvious step, but it means that users must be identifiable and there should be no sharing of passwords or other authentication methods for any services. If the authentication method is password-based, the passwords must conform to the organisation's password policy.

Device locking

Where a user is physically present to use a device, Cyber Essentials mandates that there must be an appropriate method for a user to unlock the device using a credential. This is slightly different to a password for a laptop, because although it does include that case, it also covers devices such as user-owned mobile phones and tablets that could be used to access organisational data or services.

These devices must lock themselves when not in use, and there are restrictions on the unlocking credentials. If the credentials are solely used for unlocking devices with physically present users, they may be a password, PIN or biometric credential. Passwords and PINs must be at least six characters for this purpose. If these credentials are used for any other purpose in addition to device unlocking, the full password policies of the organisation apply instead.

The unlocking credentials must also be protected against brute force attacks. Credentials should be protected by either or both of the following methods:

- Restricting the rate at which guesses can be made when incorrect credentials are presented (at most, 10 attempts per 5 minutes).
- Locking the device and permitting no further attempts to unlock after 10 incorrect attempts.

If the vendor has restricted devices so that these controls cannot be implemented, using the vendor's default settings is acceptable.

Technical control theme 3: User access control

There are two main goals in managing user access: user authentication and access minimisation. User authentication seeks to ensure that access rights to organisation data and systems are only granted to authorised users, and that anyone using these systems is who they say they are. Access minimisation aims to provide users with only the access to data and systems they need to perform their role, and nothing beyond that.

The security benefits of user authentication are obvious: users have accounts that grant access to potentially sensitive data, and only the person authorised to use that account should be using it. There are various means used to implement this, the most common being a password, although other methods such as biometric scanners exist.

User accounts should be managed: there should be a documented process for account creation and approval. All access to organisation systems and data should be through approved accounts, and all accounts should require authentication via unique credentials. No credentials should be shared. This applies to staff accounts, but also any accounts used by contractors or third parties such as suppliers and external support services.

Similarly, there should be a process for disabling or removing accounts. This should happen swiftly when an account is no longer needed (e.g. when a user leaves the organisation or moves to a role that no longer requires access to a particular system or data set). User accounts should be regularly reviewed to ensure that unnecessary accounts are dealt with. It is good practice to automatically remove or disable an account that has not been used for a significant amount of time.

Access minimisation follows the principle of 'least privilege', which is that access should only be granted to systems and data when it is necessary for the organisation's business. User accounts should only have the bare minimum of access required to perform the user's role.

Multi-factor authentication

Multi-factor authentication (MFA) is a powerful way of increasing the security of authentication methods. Instead of a single method of authentication (e.g. a password), a system requires multiple different methods – usually two – to authenticate a user's identity. Note that this goes beyond having two passwords, as multi-factor authentication requires that the additional mechanisms be of a different 'type' (or 'factor') from the original method. The available mechanism types for MFA are:

- Something you know (e.g. a password or PIN);
- Something you have (e.g. a hardware dongle or mobile phone); and
- Something you are (e.g. a fingerprint or retina scanner).

This means that having a password and a PIN would not count as MFA, as those are the same mechanism, but a password coupled with a code texted to a specific phone would fulfil the requirements for MFA. Cyber Essentials mandates specific factors that can be used to meet MFA requirements, and these are:

- A managed or enterprise-owned device;
- An application on a trusted device;
- A token that is physically separate from the system being logged in to; and
- A known or trusted account.

MFA represents a significant improvement in security over a single method. For the above example, an attacker would need to both crack a password and obtain access to (or somehow spoof) a mobile phone for the same user, which is a much more difficult task than cracking a password alone. It is best practice to use MFA wherever it is available for user authentication, and Cyber Essentials mandates its use in certain circumstances, such as Cloud-based services, administrative accounts and accounts that are accessible from the Internet.

Some methods of MFA are more secure than others. A password plus an SMS is less secure than most and should be avoided if other methods (such as a hardware dongle or an authenticator application) are available. However, a password plus an SMS is much better than a password alone.

Administrative accounts

Organisations need to take care with administrative accounts (and any other accounts with similar privileges). These accounts generally have access to system functions that go beyond what a normal user can do. For example, an administrator account may be able to disable security precautions, add or remove other accounts, and permanently delete data. An attacker that has compromised an administrator account can act with its privileges, potentially causing enormous damage to a system and greatly disrupt an organisation's business. Thus, accounts with these privileges need to be managed with additional controls over ordinary accounts.

The organisation should have policies in place that mandate that administrative accounts can only be used for administrative purposes. Administrative accounts should not be used for everyday tasks such as emailing or web browsing, and administrators should be provided with a non-privileged account for this sort of task. This keeps administrative accounts as far away from attack vectors as possible and ensures that if an administrator downloads some malware, it should only infect a non-privileged account.

Although all accounts that are no longer in use should be disabled or removed, special care needs to be taken over the removal of administrative accounts. These accounts should be reviewed regularly, and more frequently than standard accounts, and should be automatically disabled after a comparatively short amount of time not being used. It is much better to have to occasionally recreate a deleted administrator account than have it be forgotten and potentially compromised.

While this isn't a mandated part of the scheme, it is good practice to protect the identity of those who have high-privilege accounts. If an attacker can get a list of the users that have administrative access, that is useful targeting information for social engineering attacks.

Password-based authentication

Where passwords are used to authenticate a user, Cyber Essentials mandates that certain controls should be in place. Some of these, such as password length, will be familiar but ideas on best practice for passwords have changed since the start of the scheme in 2014 and the 2022 revision of Cyber Essentials introduced other requirements, including for staff education on passwords. The three areas of password controls are:

1. Technical controls to manage password quality;
2. Protection against brute-force techniques; and
3. Support for users to choose appropriate passwords.

The technical controls to manage password quality must be one or more of the following:

- The use of multi-factor authentication, as described previously.
- A minimum password length of 12 characters.

- A minimum password length of 8 characters, alongside a blacklist that disallows the use of common or compromised passwords.

As described earlier, brute-force attacks attempt to quickly guess each possible password and input the guesses to a system until the correct one is found. Cyber Essentials mandates at least one of the following measures to protect against these attacks:

- The use of multi-factor authentication.
- Lowering the rate at which authentication attempts can be made after some failures (a maximum rate of 10 guesses in 5 minutes).
- Locking an account after no more than 10 failed authentication attempts.

In addition to these technical controls, Cyber Essentials states that users should be supported in choosing unique passwords. This support includes education (see the appendix for several relevant education resources). The Cyber Essentials scheme mandates the following:

- Users should be educated in avoiding common or easily discoverable passwords, or passwords they have used elsewhere. This could include the use of password manager applications.
- Users should be encouraged to choose longer passwords. This part of the scheme does not specify a minimum number of characters or a specific method,

but suggests the use of 'three random words'[7] to create a password.

- The organisation should provide suitable secure storage for passwords (such as a locked cabinet or a password manager application) and a clear policy for when such storage can be used.

In addition to these requirements, Cyber Essentials forbids two practices that have been in common use in the past:

- The organisation should not require regular password expiry. Having passwords that expire encourages users to engage in practices such as writing passwords down and reusing passwords, which can compromise security.
- The organisation should not set password complexity requirements (such as requiring numbers, capital letters and/or punctuation marks in passwords). These requirements make passwords difficult to remember and disallow techniques such as 'three random words'.

Technical control theme 4: Malware protection

Malware describes any hostile or intrusive software that may pose a danger to an organisation. Social engineering attacks often have the goal of tricking a user into accepting harmful software. Malware is a major vector of cyber attacks and adequate protections against it are vital in the

---

[7] *https://www.ncsc.gov.uk/blog-post/the-logic-behind-three-random-words*.

modern computing environment. Malware is constantly evolving, but here are the most common types:

Viruses

Viruses have been around for long enough that the term 'virus' is often used as a shorthand for all malware, but viruses are a distinct type. A virus inserts itself into a file (often an executable) and acts when the host file is processed or run. It replicates by copying itself into other files. Viruses are defined by their infiltration and replication methods and can have various payloads, such as ransomware.

Worms

Worms are similar to viruses but don't hide in host files. They need someone to run them and are often used as part of a social engineering attack designed to trick a user into clicking a link containing a worm. Worms replicate by copying themselves and can spread quickly through a network. Worms may or may not have a payload: even a worm without a payload can cause significant damage by filling up storage space and reducing network bandwidth as it replicates. Cleaning up after a worm attack is often a long and laborious process.

Trojans

A trojan takes inspiration from the Trojan horse and needs to be invited inside. These are generally used with social engineering attacks such as phishing, and the usual trojan approach is to trick a user into installing a program that sits in the background pretending to be innocent. Most trojans don't typically inject malicious code or otherwise cause overt trouble: they commonly act as a back door into a network, allowing an attacker to access a network and the data on it. The main problem with trojans is detecting them, as a user is unlikely to report the trojan once it is installed.

Spyware

Spyware is a malware payload that monitors a user's activity. It is delivered by another method (such as a virus or trojan) and passively watches. Spyware might be used to monitor a user's web browsing habits to gather information for identity theft or to look for financial details.

Key loggers

Key loggers are a specific sort of spyware. A key logger is a program that records each press of a key a user makes and sends that information to an attacker. Key loggers can deliver any information that goes through a keyboard: usernames, passwords, typed financial reports or credit card details. Key loggers mainly affect desktop and laptop computers, but key logging programs exist for mobile phones and tablets.

Ransomware

As mentioned previously, ransomware has grown to become a significant malware threat. An attacker spreads malware across an organisation's infrastructure via an entry method (e.g. a worm) and the payload encrypts or threatens to delete most or all of the data on the system. The data is largely lost unless the organisation pays a ransom (generally in cryptocurrency) to the criminal, who may release or sell the data regardless of payment.

Bots

Bots are software applications that can be triggered remotely and can perform various tasks automatically. Bots have many legitimate uses, but they are used by attackers as a way to create 'zombie' devices that can be put to dubious ends. A bot program is a payload, and when installed on a target machine it sits quietly until it is told to do something. Criminals install and use bots in huge numbers, called botnets, as a way of gaining free computing power. Botnets can be used to do the following:

- Attack another system by flooding it with requests or messages as part of a DDoS attack.
- Mount a spam campaign by sending large numbers of emails.
- Perform click fraud: repeatedly 'watching' media to create advertising revenue or otherwise pretend to be a human user.

Rootkits

Rootkits are a very dangerous malware payload. Once installed, they can remotely control a computer, often with increased (e.g. administrator) privileges. Once a rootkit is installed, it can be very difficult to detect as the attacker can turn off or otherwise subvert anti-malware measures. Rootkits can be used to cover the existence of other malware (such as a key logger). Rootkit removal is also notoriously difficult, often leading to hardware needing to be replaced.

Countering malware

Although malware is highly diverse and uses many avenues of attack, there are several ways of protecting a system against malware attack. Anti-malware software (often called antivirus software, but it generally covers multiple malware types) is commercially available and likely to be familiar. These applications seek to identify and disable malware before it can do any harm. An organisation can also implement application controls, so that only administrators can install new software or that only software that appears on an approved whitelist can be run on a system.

Cyber Essentials mandates that all devices in scope have anti-malware protection, and each device should be protected by at least one of these mechanisms:

- Anti-malware software.
- Whitelisting.

Anti-malware software

As malware is constantly evolving, anti-malware must be kept up to date. The organisation must update the signature files (which store information on malware threats) in line with the vendor's recommendations, preferably automatically. The software must be configured to automatically scan any file accessed by users, including web pages, and look for malware. The software should also keep a list of known malicious websites and forbid access to those sites, unless there is a clear business need to access them.

Whitelisting

If an organisation allows only approved applications to run, it must obviously maintain a list of approved applications, and must vet applications appropriately before putting them on the list. The list should be reviewed regularly and unnecessary applications should be removed.

Technical control theme 5: Security update management

The fifth and final technical control theme in Cyber Essentials is security update management. Cyber security is a fast-moving discipline and new threats are appearing every day. Attackers are finding new vulnerabilities to exploit and security researchers are working equally hard on discovering security flaws before the criminals do.

Vulnerabilities can be caused many ways: sometimes the implications of a design decision haven't been thought through or sometimes a measure designed to ease testing is accidentally left in production software. However they happen, once a vulnerability has been discovered, there is a rush by security experts to remove or mitigate it. Once a solution to a vulnerability is found, it is generally published by creating a security update containing the fix. Security updates are usually released on a fixed schedule (such as once per week) but emergency updates can happen when a vulnerability is particularly dangerous or has already been exploited.

The point of security update management is to ensure that security updates are received and applied to software quickly. Security updates are no use if they have not been applied yet, and an organisation may be unaware that it is in a race with an attacker to fix a vulnerability in its system before the attacker can exploit it and cause havoc. Some software is hugely popular and is therefore the focus of many attempts to exploit any vulnerabilities it has. It is particularly important to keep operating systems and applications such as Outlook or Chrome up to date.

If one machine on a network lacks security updates, it is likely that most or all the machines on the network will also lack updates. This makes an organisation with unpatched machines a very attractive target for certain types of attacks. Ransomware is often aimed at organisations that may have lax patching policies, especially institutions that rely on a limited public purse, such as those in health or education.

Cyber Essentials requires that all software in use should be licensed and supported. This means that any software that has been abandoned by its vendor or is otherwise at the end of its life should be removed and replaced. In this context, 'supported' means still receiving security updates should any vulnerabilities be found. Supported software means that you should have access to any security patches that are created. Vendors usually support a small number of versions of their software (sometimes just one). If you use older software, you need to check that the version you have is still supported and move to a more recent version if it is not. There are few things that hackers like more than elderly, unpatched software installations. If you cannot avoid using unsupported software, you must move it to an isolated machine or network that does not have access to the Internet.

Security updates must be applied quickly. While it is not a requirement of Cyber Essentials, it is best practice to apply updates automatically wherever possible. This ensures that the updates are applied quickly and has the side benefit of eliminating the effort required for manual updating. Many software applications can be set to automatic updating, and this should cover most of an organisation's software estate.

It is good practice to review the versions of software that are installed on an organisation's machines. Some software updates give the user an option to install later, and this can lead to updates being put off for an unacceptable amount of time. An organisation may need to take measures (such as removing the delay option) if software updates are being delayed unnecessarily by users.

If a piece of software does not apply software updates automatically, manual updates should be applied quickly. Cyber Essentials recommends (but does not mandate) that all security updates should be in applied within 14 days of their release. The scheme mandates that they must be applied within 14 days in any of the following circumstances:

- The update addresses any vulnerability that has a score of seven or more from the CVSS v3 scheme.[8]

---

[8] For details of this scheme, see: *https://www.first.org/cvss/v3-1/*.

- The vendor describes any vulnerability being fixed as 'high-risk' or 'critical'.[9]
- The vendor provides no details as to how critical a vulnerability being fixed is.

In all circumstances, it is best to apply security updates as fast as is practical. These guidelines describe the maximum time that should elapse before application.

Further guidance from Cyber Essentials

The Cyber Essentials provides guidance that is not part of any technical control theme. This guidance isn't mandatory but is good practice and might help to avoid or mitigate security issues.

Back up data

An organisation should regularly back up its data. This is useful in general, but in security terms it means that if there is an attack, an organisation will have a better chance of a good recovery if its data is compromised, deleted or stolen. Cloud-based storage can be particularly useful in backing up data, although this can create security issues if not managed carefully.

At a smaller scale, data can be backed up to external drives or USB sticks. If you are using these methods, you need to remember to disconnect the backup device from the main storage when it is not performing a backup. A backup of

---

[9] If the vendor uses different terms for vulnerabilities, use the CVSS scheme as a guide.

your data will be no use to you if an attacker can find it when they find the main data store.

Asset management

Many of the technical controls within Cyber Essentials rely on an organisation tracking and managing its assets. Asset management underpins cyber security, as an organisation needs to know what IT assets it owns and who has access to them to defend against cyber attacks. Cyber Essentials does not mandate any specific method or level of asset management, but notes that asset management is a core security function and that an organisation needs to pay especial attention to managing the introduction of new assets to an organisation's IT infrastructure.

Zero trust networks

The traditional model for network security is 'castle and moat'. This is where anything outside the network is not trusted, but users and assets inside the network are trusted. An alternative to this is zero trust networks, where there is no assumption of trust within a network and authentication is controlled at a more granular level.

Zero trust networks are gaining traction within IT security. Cyber Essentials expresses no preference on the use of zero trust networks, but does note that the controls required by the scheme do not prevent an organisation from using a zero trust approach.

Part 2: Gaining cyber essentials certification

CHAPTER 4: CERTIFICATION

Implementing the controls outlined in the Cyber Essentials requirements is a valuable exercise for any organisation, but only by becoming certified can you show customers, investors, insurers and others that you are fully compliant. This is increasingly important, as nobody wishes to entrust data to an organisation that will not look after it.

Although the requirements of the scheme are relatively simple to meet, there is always a cost in time, money and organisational resources when applying a set of controls thoroughly and accurately. Failing to pass the assurance process – falling at the final fence – will increase this cost, so it is sensible to be familiar with how the process will be carried out before starting it. This will improve your chance of succeeding first time.

Note that although the Cyber Essentials scheme was created by a branch of the UK government and is required for certain UK-based contracts, it is not a solely UK-centric creation. Organisations outside the UK are welcome to apply for certification and the approach to security that Cyber Essentials mandates can benefit any organisation that uses IT.

The Cyber Essentials requirements are maintained by the National Cyber Security Centre (NCSC), while certification is the responsibility of NCSC's certification partner, the IASME Consortium. The steps to gaining Cyber Essentials certification (at the basic level, rather than Plus) are as follows:

1. Define the scope of what you wish to be certified. We will go over the details of this later.
2. Download the self-assessment questionnaire (SAQ) from the IASME website,[10] which is available in English and Welsh.
3. Prepare for certification and document the controls you have. If you are unsure whether you will be certified or need some help implementing the Cyber Essentials controls, you may wish to gain some help from an outside source. There are resources for this listed in the appendix to this book.
4. Choose a certification body.
5. Pay a fee (at the time of writing, a few hundred pounds, depending on organisation size) to the certification body to register for assessment.
6. Once you have paid the fee, you have six months to fill in the questionnaire and make any necessary changes so that your answers will be acceptable.
7. Submit your SAQ answers to the certification body. A member of your board will have to certify that the answers are true and accurate.
8. If you pass, you will be granted certification for 12 months. You may also receive free insurance (see below). If you fail, you will receive feedback on the items that caused the failure.

---

[10] *https://iasme.co.uk/cyber-essentials/free-download-of-cyber-essentials-self-assessment-questions/*.

Free insurance is, under certain circumstances, available to organisations that have Cyber Essentials certification (at the basic or Plus level). To gain this insurance, the organisation must be domiciled in the UK and have an annual turnover of less than £20 million. The insurance is designed to cover smaller incidents and has a limit of indemnity of £25,000 at the time of writing. For more details, see the IASME website.[11]

It is important to be truthful when answering the self-assessment questionnaire. The Cyber Essentials scheme is there to benefit your organisation, and any security it provides will be an illusion if your answers are less than candid. Also, any discrepancies are likely to be discovered. If (or more likely, *when*) you are the victim of a security breach, various regulators will be very interested in any security controls you said you had but in fact did not. Any gaps are also likely to be discovered if you apply for further accreditation, such as Cyber Essentials Plus or ISO 27001.

Scoping for certification

As mentioned earlier, the first step in implementing Cyber Essentials is to determine which parts of your IT infrastructure are in scope. The scoping process considers both which parts of your organisation and IT infrastructure you are applying controls to, and what sort of devices and services those controls will be applied to.

If it is at all practical, you should consider your entire organisation and IT estate to be in scope. Cyber security is

---

[11] *https://iasme.co.uk/cyber-essentials/cyber-liability-insurance/*.

only as strong as its weakest point, and employing the Cyber Essentials controls across your entire organisation will give you the most benefit and the least chance of a security breach. Note that if you are successful in applying for Cyber Essentials certification, the certificate will only cover the parts of your organisation that are in scope. Customers or other stakeholders may find it suspicious that you did not certify your whole organisation.

If your scope only covers a sub-set of your organisation, you must clearly define the scope boundary around the part of your IT infrastructure used to handle sensitive data. This means working out who is managing this infrastructure (whether that is the organisation as a whole or a particular business unit), where is it physically located and the boundary of the network. You must also ensure that the sub-set is not compromised by its connections to other parts of your organisation.

Scopes must be defined in terms of the networks they cover. Scopes can be defined as named networks only excluding all other networks or whole organisation excluding networks.

Some good examples of scoping:

- The whole company excluding an isolated scientific network.
- The whole UK company excluding workshop, development, test and guest networks.

Some bad examples of scoping:

- The corporate environment.

- All systems and IT infrastructure involved in the digital processing of customer data.
- The marketing team.

The 2022 update to Cyber Essentials changed the requirements for the devices that should be in scope. There are some exceptions, but in general, the following must be in scope if they are within the organisational boundary that you wish to certify:

- Any devices that can accept incoming network connections from untrusted Internet-connected hosts.
- Any devices that can open user-initiated outbound connections to any devices on the Internet.
- Any devices that can control the flow of data between other in-scope devices and the Internet.

In short, Internet-connected devices (with a few exceptions detailed below) must be in scope, and these are likely to be the majority of an organisation's devices. Note that the organisational scope must not be manipulated to exclude end-user devices. No scope that excludes end-user devices will be acceptable to the certification body. In-scope items should include:

- Desktop PCs;
- Laptops;
- Tablets;
- Wireless access points;
- Email, web and application servers; and
- Boundary firewalls and routers.

Removable storage devices are out of scope.

The rise of remote working and bring your own device (BYOD) mean that an organisation must protect devices that are outside of a traditional security boundary. Remote working devices may be exposed to threats such as theft and unsecured network attacks, and the risks to them need to be managed. Devices that are owned by staff are generally in scope if they can access data or services provided by the organisation. Such devices should be considered in scope unless any of the following apply:

- The device is a router owned by the user or supplied by the user's ISP.

- The device cannot communicate with other devices via the Internet.

- The device is only used for native voice, native text and multi-factor authentication purposes (e.g. a mobile phone with an authenticator application but no other connection to organisational data or services).

Applying controls to BYOD equipment is challenging because the devices don't belong to the organisation and, not being centrally purchased, are likely to vary a great deal. You may need to implement some central control of these devices, perhaps inspecting them to see that they are appropriately secured. NCSC has some guidance on the subject.[12]

Devices owned by other companies (such as those used by contractors) are not in scope, although if they are accessing company information, Cyber Essentials controls should be applied to them.

Externally managed services and scope

Cyber Essentials has rules for externally-managed services (including Cloud-based services). If the organisation's data or services are hosted externally in the Cloud, these must be in scope. The organisation is responsible for defining the controls that must be present and ensuring the Cyber Essentials controls are implemented correctly. It may be the responsibility of the service provider or the organisation (or both) to implement the controls. The Cyber Essentials infrastructure requirements document details who is responsible for control implementation in various scenarios (e.g. SaaS malware protection).

Whoever is implementing the controls on externally managed services, it is the organisation that must oversee the process and obtain the evidence that the controls have been implemented properly. In the cases where the controls are implemented by the service provider, certifications or attested compliance with relevant frameworks (such as ISO 27001 or PCI DSS) may be used as evidence, or the service provider may produce documentation to show how they have implemented the controls. This may include

---

[12] *https://www.ncsc.gov.uk/collection/device-security-guidance/bring-your-own-device*.

contractual documentation or a description of the service provider's security model.

Non-Cloud-based external services (such as remote administration) should generally be in scope, but you may not be able to attest that all the necessary controls are present, in which case you can exclude those services from the scope. If they are included, you should be able to present evidence that the service provider is complying with the relevant controls. Again, security certifications may be used as evidence.

Web-based commercial applications (developed outside the organisation) used by the organisation are in scope, apart from custom components for web applications, which are not. The main difference when certifying controls on web applications is evidence. Rather than evidence based on standards or contractual documentation, web-based commercial applications should be developed and tested in line with security best practice. You should provide evidence of these (e.g. the software author's development methodology or a recent penetration test report). Help with conducting penetration tests can be found in the appendix.

Cyber Essentials checklist

When you have defined and documented your scope, have implemented the controls and are confident that your organisation meets the requirements of the Cyber Essentials technical control themes, then you are ready to try for certification.

You may not be sure whether your organisation is ready for certification. Here is a simple checklist to help you work out whether you are properly prepared. Note that this is not a guarantee of readiness and is a 'back of an envelope' exercise. If you want to be more certain that you are ready for certification, you may wish to use a gap analysis tool (you can find a link to one of these in the appendix).

1. Are all devices in scope protected by firewalls (whether a hardware device or a software-based equivalent)?
2. Are your organisation's firewalls actively managed to ensure that rules are up to date and that only authenticated incoming connections are allowed?
3. Do all devices in scope have actively managed configurations?
4. Are all unnecessary accounts, applications and services disabled or removed from devices in scope?
5. Do your organisation's authentication methods (e.g. passwords and multi-factor authentication) match the requirements set out in the Cyber Essentials scheme?
6. Does your organisation follow the principle of least privilege, where staff only have access to the data and services necessary for their role?
7. Are all of your organisation's in-scope devices protected from malware threats by an appropriate measure (anti-malware software or whitelisting)?
8. Is your anti-malware software set to update as the vendor recommends, scan files automatically and warn you about accessing malicious websites?

9. Is all the software used by your organisation licensed and supported by its vendor (i.e. still receiving security updates)?
10. Does your organisation apply security updates to its software automatically, or at least within 14 days?

You should be able to answer 'yes' to all these questions before you apply for certification to Cyber Essentials. If you are missing just a few answers, you may be nearly there, but if most of your answers are 'no', you have work to do implementing controls before you're ready for certification.

Cyber Essentials certification process

The certification process for Cyber Essentials is governed and accredited by the IASME consortium. When you are ready to go ahead with certification you must choose a certification body. Each of these organisations has been vetted by IASME and needs to have at least one member of staff with appropriate assessment and quality assurance qualifications who can measure organisations against the Cyber Essentials requirements.

Choosing a certification body is straightforward. IASME maintains a list of certification bodies,[13] and there is more information on these in the appendix. Note that not every certification body will be able to cover Cyber Essentials Plus, so it may be in your interest to choose a body that can perform both levels of certification, such as IT Governance. As certification needs to be reissued annually, there is a

---

[13] *https://iasme.co.uk/certification-bodies/*.

benefit in finding a certification body that you can establish an ongoing relationship with, which reinforces the point of finding a body that can perform both levels of assessment.

The certification process for Cyber Essentials is equally simple. Your certification body will supply you with a copy of the Cyber Essentials self-assessment questionnaire (SAQ). Your organisation should review the SAQ and ensure that your information security controls are in line with the questionnaire. Your certification body will be able to advise you if you have questions or doubts. Once you feel confident that your measures are appropriate, an appropriate person at your company (e.g. CISO, CIO, CTO or an IT manager in a smaller organisation) will have to fill out the SAQ as the first step toward verifying your claim.

Once it has been completed, a declaration must be signed, stating that you comply with the requirements of the scheme and that your responses to the questionnaire are accurate. The signature must come from the business owner, the chief executive officer (CEO), somebody at board level or somebody in a similar role. Your organisation will not be certified without a signature from an appropriate person. If one person has both filled out the questionnaire and signed it, it may be wise to ask another member of staff or the certification body to review the document to verify that their conclusions are correct.

The Cyber Essentials questionnaire covers everything in the requirements, and questions can be broken up into the following groups. Note that certification bodies can add further questions if they are unclear on your security posture. The sections in the questionnaire are as follows:

- Your Company
- Scope of Assessment
- Insurance
- Boundary Firewalls and Internet Gateways
- Secure Configuration
  - Device Locking
- Security Update Management
- User Access Control
  - Administrative Accounts
  - Password-Based Authentication
- Malware Protection

Fortunately, it isn't necessary to answer every question perfectly – some are there simply to establish that your organisation is taking a stance on cyber security that broadly conforms with Cyber Essentials. That said, some questions focus on aspects of security that are essential for certification because they are direct requirements.

After the questionnaire has been signed, you pass it to the certification body, who will review it and decide whether your organisation has met the requirements of the scheme. They will firstly verify that you have identified the scope to be certified and that this scope is valid. They will then look at whether you have understood and complied with the requirements of the scheme. Additionally, they will be able to let you know whether you have a reasonable prospect of passing Cyber Essentials Plus.

If they are satisfied after this verification process, then they can award you the Cyber Essentials certificate and your certification will be recorded by IASME. You can also then apply for cyber security insurance, if you are eligible.

If you are unhappy with the result handed down by the certification body and feel that you have been unfairly denied a certificate, then you can go to IASME – although it may be best to ask the certification body to reconsider matters first. IASME is required to arbitrate between certification bodies and their clients regarding the results of the certification process.

Getting certified – Cyber Essentials Plus

Although we have mentioned this before, it is worth repeating: certification to Cyber Essentials Plus does not imply a higher level of security or compliance with the scheme than the basic Cyber Essentials certification. The two levels of Cyber Essentials certification share the same controls and restrictions, to the point that they use the same document to define these features. Cyber Essentials Plus offers a more robust and rigorous check that your IT infrastructure is secure. The two levels of the scheme are looking for the same thing, but Cyber Essentials Plus is looking harder.

As certification to Cyber Essentials Plus has a more involved testing process, it represents a deeper commitment to cyber security and may grant you an edge over your competitors. The range of UK government contracts that you can bid for may also widen if you have this certification.

Not all certification bodies can provide Plus-level testing. Certification bodies for Plus must meet the stringent requirements set by IASME, which include having the technical capability to perform vulnerability scans. The certification body must also, rather than having to simply verify the self-assessment process your organisation has performed, read documents and thoroughly check whether the solutions you have put in place comply with the control requirements. This requires suitably vetted staff with auditing experience. If you intend to achieve Plus certification, make sure your certification body has the experience and staff necessary to perform the checks.

The most significant additional step in Plus testing is that the certification body will carry out an internal vulnerability test on your IT infrastructure. Their aim will be to find out whether your self-assessment questionnaire answers were accurate and if the Cyber Essentials controls have been properly implemented. They will also want to see that known vulnerabilities have been addressed.

While checking that individual controls have been implemented correctly, the certification body will perform a vulnerability scan, looking for holes in your defences. This test of your preparedness will include all Internet gateways, and servers providing services directly Internet-based users within the scope you have defined. For organisations with significant numbers of remote workers or using a BYOD model, it is important to note that they will also need to test a sample of user devices that are representative of the device types used in your operation, so it is likely that you will need some remote and/or mobile devices to test.

If the certification body is satisfied that you meet the requirements of the scheme after this more thorough level of testing, they will then award you the Cyber Essentials

Plus certificate. The independent testing regime means that your customers can be more certain that you are your security posture has a robust foundation and can protect their data from basic level threats.

If you have a small number of nonconformities, your certification body may give you the opportunity to fix them before they make their final assessment. This is cheaper and more efficient than failing the certification process and starting again later.

As before, if you are unhappy with the results of your assessment you can approach IASME for arbitration between your organisation and the certification body, although an appeal to your certifying body may be sufficient to have the decision reconsidered. Note that if your organisation fails the Plus level of Cyber Essentials, you will also lose your basic certification.

After the assessment

You will have to be recertified annually to keep your certification for both the basic and Plus levels of the Cyber Essentials scheme. This will be necessary to keep to the requirements of certain contracts and to bid for others. It is possible that some customers and potential customers might insist that you recertify more frequently, although the UK government has not suggested this will be something they require yet.

The requirement for annual recertification is because the assessment (whether the basic level self-assessment or the assessment and vulnerability scans of Plus) provides a picture of your cyber security at a particular time and makes no statement about the continued viability of your approach. You may have plans to add devices to your infrastructure or change your information security policies, and in doing so you may inadvertently stop complying with the Cyber Essentials requirements. Equally, new threats to your security might emerge that warrant new requirements in Cyber Essentials. If certification lasted indefinitely, it would remain valid without a response to these threats. You might feel protected while your organisation had a security vulnerability that could lead to a breach.

When your organisation has been certified and has perhaps won the desired contracts, it may be tempting to turn away from cyber security to focus on other business areas. It would be a mistake to let your cyber security slide, however, as there is a great deal of value in watchfully maintaining the security measures you have put in place.

The average organisational cost of a data breach has been consistently going up. The worldwide average cost of a data breach has risen to $4.35 million.[14] Keeping up your basic cyber protection could quite literally save your organisation.

Many organisations will be drawn to Cyber Essentials certification to gain access to UK central and local government contracts. However, the main benefit of the scheme is to improve your security. Apart from avoiding and mitigating breaches, staying on top of cyber security can also make a difference to bringing in new business and retaining existing customers. Organisations and customers are increasingly likely to check up on the security

---

[14] *https://www.ibm.com/downloads/cas/3R8N1DZJ*.

capabilities of their partners. If they ask to look at your arrangements, it is best to be ready – it could mean the difference between gaining and losing a contract.

As your organisation grows and flourishes, it will become a more tempting target for cyber attack. While Cyber Essentials can put your organisation on an excellent security foundation, it represents the beginning of a journey. Once your organisation is more comfortable with the requirements of the scheme, you may wish to continue your information security journey with measures that go beyond the basic. You may wish to look into more robust schemes and standards, such as ISO 27001 (which is a general framework for information security) or the PCI DSS (which is a standard that addresses payment card security).

APPENDIX 1: FURTHER ASSISTANCE

Practical help and consultancy

Although implementing Cyber Essentials controls can be straightforward for staff experienced in cyber security, it is understandable that some organisations will want help understanding what the controls mean for their organisation and how to put them in place. Perhaps your organisation lacks security expertise, perhaps it is very large or perhaps you must deal with a diverse and confusing IT estate due to business processes, corporate mergers or other factors. It might be that your security staff can't currently be spared for a governance and compliance project.

If this is the case, bringing in outside assistance might be the right move. IT Governance can offer consulting services from experts in the field, helping your business to get certified more quickly and without undue hassle.[15]

You can also consult NCSC's own security professional accreditation scheme, Certified Cyber Professional assured service (CCP),[16] to find a reputable consultant as recommended by the UK government. This scheme identifies appropriate individuals in the cyber security sector who can help to put the requirements of Cyber

---

[15] *https://www.itgovernance.co.uk/cyber-essentials-scheme.*

[16] *https://www.ncsc.gov.uk/information/certified-cyber-professional-assured-service.*

Essentials in place. This is not to say that those who are recognised by CCP are the best or most suitable for your organisation, but the service means that any recognised consultant is assured to have a base level of competence in security matters.

Useful documents and further information

The Cyber Essentials scheme was based on several security principles. These principles are relevant beyond the scheme, when designing and maintaining secure systems and safeguarding data. There are several documents that expand upon the principles of Cyber Essentials.

NCSC, the creator of Cyber Essentials, has a great deal of useful guidance on its website.[17] Two of the guidance documents that you may find useful are the "Small Business Guide: Cyber Security"[18] and "10 Steps to Cyber Security".[19]

The "Small Business Guide: Cyber Security" lays out a five-step plan for smaller businesses to improve their information security posture. The guide also details some further actions once the five steps have been completed, and provides some useful resources. Some of the measures described in the guide overlap with the Cyber Essentials scheme, but a small business interested in cyber security would be well advised to take the steps outlined in both.

---

[17] *https://www.ncsc.gov.uk/section/advice-guidance/all-topics*.

[18] *https://www.ncsc.gov.uk/collection/small-business-guide*.

[19] *https://www.ncsc.gov.uk/collection/10-steps*.

The "10 Steps to Cyber Security" guide was used as a source for creating the Cyber Essentials scheme, but it goes into greater detail about cyber threats and incident management, as well as presenting real-world examples of damaging cyber attacks. Sharing this guidance around may be a good way to convince managers within your organisation of the need to act on cyber security or to continually monitor it following certification.

The NCSC's website also hosts its Board Toolkit.[20] This is intended to promote and develop cyber security literacy in the top management of an organisation. Cyber security can only be achieved with the agreement and support of an organisation's board, and the toolkit seeks to increase knowledge of security matters among the most senior managers. The toolkit will not make the board members technical experts, but it will help them gain the knowledge that they need to effectively protect the organisation.

The next step – cyber security standards

Cyber Essentials will cover the basics of your cyber security posture, but if you are looking to implement stronger protection and to reap the rewards of being certified to an internationally recognised standard, there are other options. This is especially true if your organisation is large or presents a tempting target for an attacker. The latter might be true if you handle valuable or sensitive data, such as financial transactions or the personal details of many thousands of people. For companies looking to take the

---

[20] *https://www.ncsc.gov.uk/blog-post/cyber-literacy-for-the-boardroom*.

next step in protecting their information and that of their customers, ISO/IEC 27001:2022 (often shortened to ISO 27001) is a good choice.

ISO 27001 gives you all the tools you need to create an information security management system (ISMS) that can form the basis of a rigorous and rugged security posture. As the recognised international standard for cyber security, it will also send a strong message to your customers and potential customers that you are taking their security seriously. Due to its wide-ranging remit covering people, processes and technology, it ensures appropriate technical and organisational controls are in place to combat cyber threats and is highly recommended by IT Governance. ISO 27001 is supported by ISO 27002, which gives practical guidance on how to implement the controls.

Alternatively, the IASME Consortium has created its own standard: IASME Cyber Assurance.[21] This does not have the same level of recognition enjoyed by the ISO standards, but it may be a good choice for smaller organisations as it has been written with their needs and capabilities in mind.

Implementing any of these standards will also ensure that you comply with the requirements of Cyber Essentials. Conversely, becoming compliant with the Cyber Essentials scheme will give you a firm basis from which to comply with more demanding standards, and work towards a formidable 'defence in depth' approach to cyber security.

The Cyber Essentials scheme predates the adoption of the EU General Data Protection Regulation (GDPR). Although

---

[21] *https://iasme.co.uk/iasme-cyber-assurance/*.

the two are not directly related, and the GDPR is a law rather than a standard, they share a few goals. The GDPR is concerned with privacy and how personal data is handled by organisations. The UK leaving the EU prompted some changes to legislation, but the GDPR is still substantially in force in the UK, and any organisation based in (or handling data from) the UK or EU is very likely to have obligations under the Regulation.

The full requirements of the GDPR are too complex to detail here, but among them is a need for organisations to adequately protect personal data that the organisation holds. Even the smallest organisation is likely to hold some personal data – staff names and addresses, for example – and significant fines are possible for organisations that fail to look after it. Infringing the GDPR carries a maximum fine of €20 million (around £17 million), or 4% of an organisation's annual turnover, whichever is greater. These very large fines tend to be aimed at large organisations that have flagrantly ignored their responsibilities to look after data, but nonetheless it is worth noting that the GDPR is a piece of legislation with significant financial teeth.

Complying with the GDPR can be a significant effort but implementing with the controls specified by the Cyber Essentials scheme puts an organisation on a good footing, A Cyber Essentials compliant organisation has a head start on the data security aspects of the GDPR. That said, there is significantly more to the GDPR than data security. The Cyber Essentials scheme does not cover such aspects as data minimisation and lawful data processing.

Appendix 1: Further assistance

Staff training

As cyber attacks become more prevalent, basic precautions against them are becoming better known. Many people now recognise common cyber attacks when they see them and are aware that, for example, they should not open attachments from unsolicited emails. That said, it is a good idea to define some knowledge and processes and make your staff aware of them, ensuring that there is a consistent level of knowledge and good practice across your entire organisation. A truly effective security approach must focus on people as well as processes and technology. The best technical solutions cannot protect you if your staff cannot use them or can be easily convinced to bypass protections.

The Cyber Essentials scheme already has detailed requirements for passwords for all users, including staff education. Since you must communicate this information to staff anyway, this may be a good time to share other information.

In "10 Steps to Cyber Security", the UK government suggests the following as a start when educating your staff:

- *"Encourage senior leaders to lead by example."* Security measures should not be onerous, but they often require a little effort or inconvenience. If the senior managers can be seen to be supporting and engaging with security controls, it is more likely that other staff will accept those controls without issue.
- *"Build effective dialogue with your staff."* This will help your staff accept and cooperate with security

controls. It is also important for matters such as the principle of least privilege. It is important to understand what your staff is doing and what data they need access to in order to create security procedures that limit the amount of data and privileges that staff can access to the minimum for their job roles. It is also important not to stigmatise security mistakes, which could lead to staff failing to report security issues in the future.

- *"Consider running awareness campaigns."* You should train your staff to be aware of and able to spot several common kinds of attack (e.g. phishing), which should grant a base level of knowledge. The campaign should be seen as coming from senior management and you should focus on what staff can do to help, rather than the consequences of a breach.

- *"Tailor cyber security training to address your needs."* Make sure you understand the needs of your organisation and its staff before you start any training programme. Some threats are common across all IT, but there may be specific dangers that your industry or organisation faces, which need to be emphasised to staff. Let your staff know how this training will benefit them and deliver it in small chunks.

Here are some suggested key outcomes for a cyber security training programme. Staff should:

- Be aware of their cyber security responsibilities;
- Use a strong password;

- Be on the lookout for phishing and other social engineering attacks;
- Know not to send emails or other communications that could damage the company's reputation or jeopardise business; and
- Know never to open spam or other suspicious emails.

Cyber resilience

Cyber security is your first line of defence, but with the proliferation of cyber attacks, one will inevitably get through the security measures of even the most well-protected organisation. As a result, cyber resilience – defined as an organisation's ability to recover from a successful attack and return to business as usual – is an important consideration.

Cyber resilience includes the discipline of cyber security – the measures outlined in the Cyber Essentials scheme, for example – but there is more to it. It is also concerned with:

- ***Detecting attacks*:** Organisations need to be able to detect attacks quickly so that they can rapidly respond and minimise the damage. This includes having systems in place to look for suspicious activity and training staff to spot the signs of an attack.
- ***Responding to attacks*:** Once an attack has been detected, organisations need to have a plan in place for how to respond to minimise the damage. This should include who to contact and what steps to take.

- ***Recovering from attacks*:** Once an attack has been successfully dealt with, organisations need to be able to recover their systems and data. This includes having backups in place and a plan for how to restore systems.

Putting a complex strategy like this in place requires a significant investment of time, resources and intellectual capital, so you should seek out expert guidance and dedicated cyber resilience resources before deciding whether to take this step.

APPENDIX 2: IT GOVERNANCE RESOURCES

IT Governance offers several unique solutions to help you meet the requirements of the Cyber Essentials scheme at a pace and for a budget that suits you.

As an IASME-accredited certification body, IT Governance can help you achieve certification to either Cyber Essentials or Cyber Essentials Plus.

Certification only

- You read the requirements, implement them, then complete and submit the SAQ.
- We then assess the questionnaire (and perform an internal scan and assessment for CE Plus) and issue the certificate subject to compliance.
- *https://www.itgovernance.co.uk/shop/product/cyber-essentials-certification*

Get A Little Help

- We teach you what to do and give you the tools. We give you two hours of consultancy support and you implement any changes, then complete and submit the SAQ.
- We then assess the questionnaire (and perform an internal scan and onsite assessment for CE Plus) and issue the certificate, subject to compliance.

- *https://www.itgovernance.co.uk/shop/product/cyber-essentials-get-a-little-help*

Get A Lot Of Help

- We show you what to do via up to seven hours of consultancy, and help you complete and submit the SAQ.
- We then assess the questionnaire (and conduct an external scan for CE, and an internal scan and assessment for CE Plus) and issue the certificate subject to compliance.
- *https://www.itgovernance.co.uk/shop/product/cyber-essentials-get-a-lot-of-help*

Cyber Essentials Plus Health Check

- We come on site and assess your current security approach for compliance with Cyber Essentials and your likelihood of passing Cyber Essentials Plus certification.
- Also includes a gap analysis against the five technical control themes, guidance on scoping and a roadmap report for what you need to do for compliance.
- *https://www.itgovernance.co.uk/shop/product/cyber-essentials-plus-health-check*

Visit *www.itgovernance.co.uk/ces-certification* for more information.

Penetration testing

The best way to find out whether your infrastructure is secure is to test it. Penetration testers are ethical hackers who will probe networks to discover security flaws before a criminal does. The vulnerability scans that are required for Cyber Essentials Plus certification are a form of penetration test.

Penetration testing is a broad area, and the costs and methods involved depend on what is being tested, the issues being looked for and how much effort the organisation is willing to put into the matter.

IT Governance offers a broad variety of penetration testing services.

Visit *www.itgovernance.co.uk/penetration-testing* for more information.

Gap analysis

If you suspect that your organisation is not quite ready for Cyber Essentials certification, you may find a gap analysis tool helpful. These are available for a very modest price and will help you go through the control requirements, logging where your organisation stands and flagging the controls that you need to work on. A gap analysis tool will also help you monitor your progress in control implementation, giving you assurance that you will know when you are compliant with requirements.

Visit *www.itgovernance.co.uk/shop/product/cyber-essentials-gap-analysis-tool* for more information.

GRC eLearning courses

GRC eLearning provides staff awareness courses that cover cyber security and data protection issues, among other topics. Some of these courses will cover the requirements for staff education on cyber security in the Cyber Essentials scheme.

Cyber Security Staff Awareness E-learning Course

- A thorough grounding in cyber security matters for non-specialist staff.
- Delivered online, only 45 minutes long and includes a test.
- *https://www.grcelearning.com/product/cyber-security-staff-awareness-e-learning-course*.

Visit *www.grcelearning.com/category/courses* for more information.

FURTHER READING

IT Governance Publishing (ITGP) is the world's leading publisher for governance and compliance. Our industry-leading pocket guides, books, training resources and toolkits are written by real-world practitioners and thought leaders. They are used globally by audiences of all levels, from students to C-suite executives.

Our high-quality publications cover all IT governance, risk and compliance frameworks and are available in a range of formats. This ensures our customers can access the information they need in the way they need it.

Our other publications about cyber security include:

- *ISO/IEC 27001:2022 – An introduction to information security and the ISMS standard* by Steve Watkins,
  www.itgovernancepublishing.co.uk/product/iso-iec-27001-2022
- *The Cyber Security Handbook – Prepare for, respond to and recover from cyber attacks* by Alan Calder,
  www.itgovernancepublishing.co.uk/product/the-cyber-security-handbook-prepare-for-respond-to-and-recover-from-cyber-attacks
- *IT Governance – An international guide to data security and ISO 27001/ISO 27002, Eighth edition* by Alan Calder and Steve Watkins,

www.itgovernancepublishing.co.uk/product/it-governance-an-international-guide-to-data-security-and-iso-27001-iso-27002-eighth-edition

For more information on ITGP and branded publishing services, and to view our full list of publications, visit *www.itgovernancepublishing.co.uk*.

To receive regular updates from ITGP, including information on new publications in your area(s) of interest, sign up for our newsletter at *www.itgovernancepublishing.co.uk/topic/newsletter*.

Branded publishing

Through our branded publishing service, you can customise ITGP publications with your company's branding.

Find out more at *www.itgovernancepublishing.co.uk/topic/branded-publishing-services*.

Related services

ITGP is part of GRC International Group, which offers a comprehensive range of complementary products and services to help organisations meet their objectives.

For a full range of resources on Cyber Essentials visit *www.itgovernance.co.uk/cyber-essentials-scheme*.

Training services

The IT Governance training programme is built on our extensive practical experience designing and implementing management systems based on ISO standards, best practice and regulations.

Our courses help attendees develop practical skills and comply with contractual and regulatory requirements. They also support career development via recognised qualifications.

Learn more about our training courses and view the full course catalogue at *www.itgovernance.co.uk/training*.

Professional services and consultancy

We are a leading global consultancy of IT governance, risk management and compliance solutions. We advise businesses around the world on their most critical issues and present cost-saving and risk-reducing solutions based on international best practice and frameworks.

We offer a wide range of delivery methods to suit all budgets, timescales and preferred project approaches.

Find out how our consultancy services can help your organisation at *www.itgovernance.co.uk/consulting*.

Industry news

Want to stay up to date with the latest developments and resources in the IT governance and compliance market? Subscribe to our Weekly Round-up newsletter and we will send you mobile-friendly emails with fresh news and features about your preferred areas of interest, as well as unmissable offers and free resources to help you successfully start your projects. *www.itgovernance.co.uk/weekly-round-up*.

EU for product safety is Stephen Evans, The Mill Enterprise Hub, Stagreenan, Drogheda, Co. Louth, A92 CD3D, Ireland. (servicecentre@itgovernance.eu)